Water challenges are facing communities and regions across the United States, impacting millions of lives and costing billions of dollars in damages. Recent events, including record-breaking drought in the West, severe flooding in the Southeast, and the water-quality crisis in Flint, MI, have elevated a national dialogue on the state of our Nation's water resources and infrastructure. This dialogue is increasingly important as a growing population and changing climate continue to exacerbate water challenges. On March 22, 2016—World Water Day—the Obama Administration hosted the first-ever White House Water Summit to shine a spotlight on the importance of cross-cutting, creative solutions to solving the water problems of today, as well as to highlight the innovative strategies that will catalyze change across the ways in which we use, conserve, protect, and think about water in the years to come. As part of the Summit, the Administration called on institutions and organizations from all sectors to make new commitments to build a sustainable water future in the United States. In response, institutions and organizations made the following commitments, as reported and described by respondents.

Contents

New Steps Being Taken by the Administration

Today, the White House is announcing new steps to help build a sustainable water future in the United States. These steps include:

- **Building national capabilities for long-term drought resilience.** Drought routinely affects millions of Americans and poses a serious and growing threat to the security of communities nationwide. While drought has recently been particularly detrimental to Western states, serious drought issues can affect nearly every region of the country; in 2012, drought covered more than 65% of the United States. Drought presents risks to the security of the U.S. food supply and integrity of critical infrastructures, causes extensive economic impacts, increases energy costs, and adversely impacts health in many ways. The impacts of climate change are expected to increase the frequency, intensity, and duration of droughts in many regions. That's why today, President Obama is issuing a Presidential Memorandum on Building National Capabilities for Long-Term Drought Resilience. The Memorandum lays out six drought-resilience goals and corresponding actions, and permanently establishes the National Drought Resilience Partnership (NDRP) as an interagency task force responsible for coordinating execution of these actions. In addition, building on previous drought-response efforts, the Administration is releasing the NDRP's Long-Term Drought Resilience Federal Action Plan. The Action Plan describes specific activities that Federal agencies will take — within existing resources and authorities and working with State, regional, tribal, and local partners — to build national drought-resilience capabilities in accordance with the goals and actions of the Presidential Memorandum. These actions build on previous efforts of the Administration in responding to drought and are responsive to input received during engagement with drought stakeholders, which called for shifting focus from responding to the effects of drought toward supporting coordinated, community-level resilience and preparedness to adapt to drought impacts.

  In conjunction with today's Presidential Memorandum and NDRP Action Plan, Federal agencies are announcing new efforts to enhance long-term drought resilience:
  - Improving drought monitoring and forecasting. In 2016, as part of the NDRP Action Plan, the **U.S. Department of Agriculture (USDA)** and **NOAA** will expand the U.S. Drought Monitor — a vital tool for guiding response to drought emergencies — to include the U.S. Virgin Islands and the U.S. Affiliated Pacific Islands.
  - Improving rural access to drinking water. As part of the NDRP Action Plan, **USDA** will work with States and tribes to identify rural communities most at risk for compromised drinking-water supplies as a result of drought. Additionally, USDA will make inclusion of drought impacts in emergency-response plans a condition of funding for new water and waste infrastructure projects, and will train technical-assistance providers as needed to support communities in meeting this requirement.
  - Improving the coordination and integration of Federal programs. In 2016, as part of the NDRP Action Plan, **USDA** and the **U.S. Bureau of Reclamation (USBR)** will extend the successful practices of cross-program coordination for USBR WaterSMART and USDA Environmental Quality Incentives Program (EQIP) water-efficiency grants currently underway in California to other basins suffering from or at risk for drought.

1

- Announcing a new public-private collaboration around drought research. The **U.S. Geological Survey (USGS)** National Climate Change and Wildlife Science Center is announcing a new partnership with the **Wildlife Conservation Society** and **The Nature Conservancy** to synthesize current understanding of the ecological impacts of drought and examine sets of management options that are relevant at the national, regional, and local levels.

- **Supporting cutting-edge research.** The Administration is announcing a number of efforts to support cutting-edge water-research projects, including:
  - Awarding nearly $35 million in grants:
    - The **National Science Foundation (NSF)** is providing $20 million through its Experimental Program to Stimulate Competitive Research (EPSCoR) to research teams who will apply a systems-based, highly integrated approach to examine impacts of extreme events. An integrated model of the watershed will be used to test management scenarios and identify strategies for maintaining infrastructure, environmental health, and drinking-water quality in the face of extreme-weather events. NSF is also providing $2 million through its Advanced Technology Education program to educate technicians for water-related and other high-technology fields that drive our Nation's economy.
    - **USDA** is awarding $8.5 million to ten institutions and organizations through its National Institute of Food and Agriculture's Water for Agriculture program, to support research into critical water problems in rural and agricultural watersheds across the United States. These grants further research and education projects focusing on sustainability, allocation, and management of water resources, as well as the treatment and safety of water sources.
    - The **Environmental Protection Agency (EPA)** is awarding $3.3 million to five institutions—the Water Environment Research Foundation; the University of Illinois at Urbana-Champaign; Utah State University; the University of Nevada, Las Vegas; and the University of California, Riverside—to fund research on the health and ecological impacts of water-conservation practices. EPA will also hold a kick-off event for $4 million recently awarded to four institutions—Public Policy Institute of California, Water Research Foundation-University of Colorado Boulder, University of Utah, and Clemson University—to fund research on potential impacts of drought and forest fire on water quality.
  - The **National Aeronautics and Space Administration (NASA)** is announcing the formation of a new, agency-wide Western Water Applications Office (WWAO), based at its Jet Propulsion Laboratory (JPL) at the California Institute of Technology in Pasadena. The WWAO will support the strategic development of key applications from satellite observations and airborne technologies to maximize their use in order to better meet the challenges of drought, flooding, declining snowpacks, and falling groundwater levels across the west. NASA is launching this effort in summer 2016.
  - The Federal agencies participating in the **National Nanotechnology Initiative** are announcing a new Nanotechnology Signature Initiative (NSI), *Water Sustainability through Nanotechnology*. The new NSI will focus on applying the unique properties of

materials—including increased surface area and reactivity—that occur at the nanoscale to increase water availability, improve water delivery and use efficiency, and enable next-generation water-monitoring systems. Participating agencies include the **Department of Energy (DOE), EPA, NASA, the National Institute of Standards and Technology (NIST), NSF,** and **USDA**.

- **Piloting promising solutions.** Testing and demonstration of new approaches to water sustainability is an essential precursor to large-scale implementation. Today, the Administration is announcing pilots of several such approaches:
    - <u>Improving weather forecasts for water-management operations.</u> This year, **NOAA, USGS,** and the **U.S. Army Corps of Engineers (Army Corps),** along with the **Sonoma County Water Agency** and other local and state partners, will launch the Lake Mendocino Forecast Informed Reservoir Operations pilot project in California's Russian River. This pilot will demonstrate ways in which improved weather forecasts can aid the decisions made by Army Corps and other water-resource managers as they balance flood and drought risks, maximize reservoir-storage potential, and minimize conflict among competing water users.
    - <u>Improving identification and monitoring of harmful algal blooms.</u> In August 2016, **NOAA,** the **University of Michigan,** and the **Monterey Bay Aquarium Research Institute** will deploy the Environmental Sample Processor (ESP) in Lake Erie for the first time. The ESP "lab-in-a-can" will be deployed autonomously to collect water samples, run molecular diagnostics, and provide water managers with data on harmful-algal toxicity in near real-time before the water reaches municipal water intakes.
    - <u>Enhancing water sensing.</u> The NOAA-funded **Alliance for Coastal Technologies, EPA, USGS** and other partners are collaborating with **XPRIZE** to create pilot opportunities to demonstrate uses of sensors from the Nutrient Sensor Challenge and the Wendy Schmidt Ocean Health pH XPRIZE in a wider variety of "real-world" conditions and settings. In 2016, these innovations will be tested and verified as components of existing operational environmental monitoring and observing systems and networks.
    - <u>Reducing water use in power plants.</u> The **DOE** Office of Fossil Energy is issuing a competitive funding opportunity for development of a 10 MW scale test facility for validating the performance of power cycles that use supercritical carbon dioxide instead of water as the working fluid—an approach with the potential to considerably reduce the water requirements of power generation.

- **Supporting water-innovation networks.** Building on the Nation's historical reputation for ingenuity, the Administration is announcing new efforts to connect researchers, technologists, and innovators across the country to accelerate solutions to priority water challenges.
    - <u>Recovering resources from wastewater.</u> **DOE, EPA, NSF,** and **USDA,** in collaboration with the **Water Environment Research Foundation,** are developing a National Water Resource Recovery Test Bed Facility network and directory, to connect those working on approaches for recovering energy and other valuable resources from wastewater with test facilities appropriate for their needs. Today, the collaboration is announcing the next step in this effort: two NSF-sponsored

3

workshops in May and June of this year to develop metrics and structure for the network.

- Supporting innovation clusters. **EPA** is committing to provide $200,000 in funding this year for its recently established Environmental Technology Innovation Clusters Program. The Cluster Program supports a network of 15 regional groupings of businesses, government, research institutions, and other organizations focused on development and deployment of technologies for addressing the Nation's water and other environmental challenges. In addition, the Cluster Program is releasing a statement from Cluster leaders recommending core actions to advance water innovation in the United States.

- **Expanding monitoring and forecasting capabilities.** Accurate, timely, and sufficient data, information, and predictions about our Nation's watersheds and water cycles are critical to informing planning and decision making at all levels. That's why the Administration is announcing new steps to expand these capabilities:
  - Releasing a new National Water Model. In June 2016, **NOAA** will release a new National Water Model that will dramatically enhance the nation's river-forecasting capabilities. The model—which relies on data from EPA and USGS and was developed by the National Center for Atmospheric Research (NCAR), funded by NOAA and NSF—will deliver forecasts for approximately 2.7 million locations, up from 4,000 locations today—a 700-fold increase in forecast density. Other institutions are already launching their own workstreams to build on the new model, including:
    - The NSF-funded **Consortium of Universities for the Advancement of Hydrologic Science, Inc. (CUAHSI)** and the **University Consortium for Geographic Information Science (UCGIS)** will work with **NOAA** and other Federal agencies to analyze the Nation's land-surface elevation—an important step in being able to add real-time flood-inundation mapping to the National Water Model. This project will be supported by computation at the CyberGIS facility at the **University of Illinois at Urbana-Champaign,** and by a seven-week Summer Institute for graduate students at NOAA's National Water Center in Alabama.
    - **Esri** and **KISTERS North America, Inc.,** in collaboration with the academic community and **NOAA**, will build on the National Water Model and the recent success of the National Flood Interoperability Experiment to develop a National Flood Model that enhances flood forecasting for the Nation. KISTERS will work with the Center for Research in Water Resources of the **University of Texas at Austin** to integrate flood-relevant data from local agencies and other sources into the National Flood Model, with a goal of launching a pilot version of the National Flood Model later this year. Esri will develop data-processing and spatial-analysis workflows to advance research into streamflow and flood-inundation forecasting, and will make the forecasts from the National Flood Model freely available online. In addition, Esri will help visualize this data through interactive online-mapping applications that will combine the forecasts with other data to identify at-risk populations and help inform decision making.
  - Advancing western water data. **NASA**'s JPL is committing to treating western water issues with the same urgency and rigor as its spaceflight projects. By integrating

hydrological observations from NASA's flagship satellite and aircraft platforms, JPL's new Western States Water Mission will provide a high-resolution picture of western water availability (snow, surface water in rivers and reservoirs, soil moisture, and groundwater) that has not been previously possible. The data will be widely accessible on the same advanced visualization platform that NASA uses for its Mars missions.

o Enabling early identification of algal blooms. **EPA, NOAA, NASA,** and **USGS** are collaborating to develop an early-warning indicator system using historical and current satellite data to detect algal blooms, which can severely impact drinking-water quality, in U.S. freshwater systems. As part of this effort, in 2015 the collaborating agencies launched the Cyanobacteria Assessment Network (CyAN) project, which will create a standard and uniform approach for early identification of algal blooms, with an initial focus on high-priority states. Today, the project is announcing that it will expand to continental coverage by 2017.

- **Improving information and tools.** The following new Federal resources and actions will help inform planning and decision-making with respect to our Nation's water resources:
  - o Memorandum to enhance Federal coordination. **Army Corps, USGS, NOAA** and **FEMA** are today renewing a Memorandum of Understanding (MOU) for Collaborative Science, Services and Tools to Support Integrated and Adaptive Water Resources Management. The MOU will increase collaboration and partnership in areas of mutual interest to address challenges, streamline processes, and share data and information (both within the Federal government and with non-Federal institutions) in order to increase efficiency and enhance service delivery. Important efforts where progress has been made include national flood inundation mapping, systems operability and data synchronization, as well as coastal and climate-related activities. An action plan will be developed this year to guide activities for the next five years under the MOU.
  - o SECURE Water Act: Report to Congress and Visualization Tool. **USBR** is releasing a new Report to Congress—as mandated by the SECURE Water Act of 2009—that provides a basin-by-basin overview of impacts to U.S. water supplies from climate change, and includes numerous potential adaptation strategies relevant to each basin. USBR will also release an interactive SECURE Water Act Visualization Tool— a web-based companion product to the Report to Congress that allows the public to interact with the data presented in the Report, and to better understand the risks that the data indicate.
  - o Hydrologic Engineering Center River Analysis System (HEC-RAS) Update. The **Army Corps** is releasing an update to its HEC-RAS engineering software, allowing the software to perform two-dimensional hydrodynamics along with integrated one- and two-dimensional modeling, and unsteady flow computations. The two-dimensional capabilities allow users to determine the timing and direction of river flow, important for evaluating environmental and stream stability issues, and for studying consequences associated with possible dam- and levee-failure scenarios. The update also includes new capabilities for modeling surface water–groundwater seepage, as well as water-quality and sediment-transport modeling enhancements.
  - o Water-resources dashboard. **NOAA** and several outside organizations are launching a shared water-resources dashboard as part of the U.S. Climate Resilience Toolkit.

This dashboard will serve as a common resource for urban planners and local officials to easily access many of the flood and drought data sets needed to support climate-adaptation planning. To connect users around the dashboard, NOAA will run a series of web sessions to explain the data on the dashboard and demonstrate how these data can be incorporated into decision making.

o Flood flow-change detection tool. The **Army Corps** is releasing a web tool that will enable users to detect nonstationarities, or significant changes, in the statistics of annual maximum daily streamflow at any USGS gage site. The tool will improve understanding of climate variability, which will in turn allow water-resources planners and engineers to better understand and project how streamflow is, and will continue to be, affected by climate change.

o Nearshore processes research. The **Army Corps** is announcing that this fall, it will release an implementation plan for addressing the research needs identified in the 2014 *Future of Nearshore Processes Research* report. The plan, which was cosponsored by the American Shore and Beach Preservation Association and is being developed by a collaboration among 30 institutions and eight Federal agencies, will integrate research from the Federal government, academia, industry, and NGOs to inform recommendations for managing water quality and other important factors in the often highly developed yet vulnerable nearshore environment.

o Report to Congress on Upper Mississippi River Restoration. This year, the **Army Corps** will release the fourth in a series of reports to Congress on the status of the Upper Mississippi River Restoration, an effort that includes five States (IA, IL, MI, MN, and WI) as well as the EPA, USGS, the USDA Natural Resources Conservation Service, the U.S. Fish and Wildlife Service, and numerous other partners and stakeholders. The report will include information on partnership among Federal and state agencies and other organizations; construction of high-performing habitat restoration, rehabilitation projects; and increased understanding through monitoring, research, and assessment, and engagement with other organizations.

- **Raising public awareness and engagement.** To give more individuals and communities across the country the opportunity to learn more about and share their thoughts on water, the Federal government will:

  o Create a new video series. In a new video series produced by **NBC Learn,** the educational arm of NBCUniversal News Group, **NSF** will explore how cutting-edge science and engineering research can transform how the country understands, designs, and uses water resources and technologies. The four-part series, which will be made freely available for public and classroom use across a variety of platforms in fall 2016, will promote public awareness of water infrastructure designs and needs, water conservation in rural and urban settings, water-treatment techniques, and water-quality issues.

  o Host a National Climate Game Jam—Water! From April 15–24, 2016, **NOAA** will host a National Climate Game Jam—Water!, bringing together youth, climate scientists, and educators at sites around the country to create new virtual and physical game prototypes that allow players to learn about climate change and water through science-based, interactive experiences. The winners of the Jam, along with the full list of game ideas and videos, will be posted online. This

Game Jam follows on an initial commitment through the Climate Education and Literacy Initiative.

Advancing Water Sustainability on All Fronts

Across the country, stakeholders in all sectors have responded to the Administration's call to action for an all-hands-on-deck effort to build a sustainable water future, including with new actions being announced today.

Managing Water for the Long Term

To reduce and mitigate the incidence and impact of water stresses on U.S. communities, it is essential to develop, implement, and normalize sustainable, integrated, long-term water-management strategies. Today, states and localities are making new commitments to lead on this front.

- The **Colorado River Basin States (CO, WY, NM, UT, CA, NV, AZ)**, the **Southern Nevada Water Authority**, the **Metropolitan Water District of Southern California**, the **Central Arizona Water Conservation District, Denver Water**, the **Upper Colorado River Commission**, and **USBR**—are committed to addressing water scarcity on the Colorado River, a critical source of water for 40 million people, businesses, and the environment in the United States and Mexico. Based on the success of first Pilot agreements, today, the coalition is announcing the launch of Phase II of a program that brings together farmers, ranchers, and tribes with municipalities and policymakers to conserve water for the long term. The program compensates water users for implementing voluntary water-conservation projects that decrease use, improving critical water-storage levels at Lakes Powell and Mead for the benefit of the entire Colorado River Basin.

- The **State of Oklahoma** is launching its Water for 2060 Initiative, a unified approach for achieving the goal set by Oklahoma Governor Fallin and the State Legislature of using education and incentives to ensure that Oklahoma's freshwater use in 2060 is at or below 2012 levels, while supporting Oklahoma's continued population growth and economic prosperity. The initiative establishes a unified approach across each major water-use sector for increased water conservation, recycling, and reuse. By establishing a plan with a 50-year outlook, the Initiative hopes to ensure that all current and future Oklahomans have access to a readily available supply of clean, safe water for many more decades to come.

- The **City of Los Angeles** is committing to capture an additional 12 billion gallons per year of stormwater for infiltration and reuse by 2025, on top of the more than 8.8 billion gallons the City captures today. Stormwater capture is a key component of the City's goal to source 50% of its water locally by 2035 and helps fulfill multiple objectives in the City's Sustainable City pLAn.

- The **City of Tucson** and the **City of Phoenix** are announcing the next step in their exchange agreement, signed in 2014, that allows Phoenix to store water allocated under the Central Arizona Project in Tucson's underground recharge facilities, which simultaneously decreases both pumping costs for Tucson and construction costs for Phoenix. Today, the Cities are announcing that over the next year, they will work together to achieve a more than five-fold increase of water stored under this agreement, resulting in storage of more than 1.6 million gallons of water, or enough to serve over 17,000 homes for a year.

- The **Santa Ana Watershed Project Authority (SAWPA)** is implementing its "One Water One Watershed" integrated regional water management 2.0 plan with a $100 million watershed program to deal with long-term drought through water-use efficiency and conjunctive-use storage of water within the groundwater basins of the Santa Ana watershed. SAWPA will share lessons learned from this program with other drought-stricken regions across the western United States.

Investing in Water Solutions

The availability of private capital is an essential component of ensuring long-term solutions to complex environmental and social challenges. Recognizing this, the Administration has launched numerous efforts and institutions—including the Clean Energy Investment Initiative, the Department of the Interior (DOI)'s Natural Resources Investment Center, EPA's Water Infrastructure and Resiliency Finance Center, and USDA's Rural Opportunity Investment Initiative—to encourage creative financing opportunities that help address these challenges while advancing economic development goals. The following private companies responded to the Administration's call to action and today are announcing new steps to invest in the Nation's water future, including nearly $4 billion in financing for water-infrastructure projects.

- The **Baton Rouge Water Company (BRWC)** recently constructed a "scavenger well couple" to achieve *in situ* separation of brackish and fresh water in an underground drinking-water aquifer. Today, BRWC is announcing that it will invest an additional $40,000 this year into a redesign of the couple's pumping equipment, to enable the Company to remove chlorides at a higher rate and extend the usefulness of the aquifer years into the future. In addition, BRWC is announcing that in 2016 and 2017, it will invest resources in another saltwater-intrusion prevention project, focusing on a different and deeper chloride-threatened groundwater drinking-water source in the local area. As part of this project, which involves **USGS**, the **Capital Area Groundwater Conservation Commission**, and **Louisiana State University**, BRWC will temporarily dedicate two of its groundwater drinking-water aquifers for development and testing of new saltwater-intrusion prevention techniques.

- **CDP** is introducing water security into its supply-chain program, supporting U.S. businesses in reducing water impacts and enhancing water security across their vast supply chains. Through CDP's supply-chain program, companies will use water data from more than 1,500 suppliers to shift $218 billion worth of corporate procurement spending to support sustainable water use.

- To help foster investment in resilient and sustainable water infrastructure, **Ceres**, the **Climate Bonds Initiative**, the **Alliance for Global Water Adaptation, CDP,** and the **World Resources Institute** are launching a Water Climate Bonds Standard to provide investors with verifiable, science-based criteria for evaluating water-related bonds and to assist issuers in the global corporate, municipal, sovereign, and supra-sovereign markets in differentiating their green-bond offerings. The **San Francisco Public Utilities Commission** expects to be the first issuer to align a forthcoming bond sale with the standard in order to finance sustainable stormwater management and wastewater projects.

- The **Municipality of Anchorage** is announcing major infrastructure improvements to advance water and energy efficiency at its new $300 million Sullivan Plant 2A power-generation facility in northeast Anchorage. Two city-owned utilities, Municipal Light & Power (ML&P) and Anchorage Water & Wastewater Utility (AWWU), are partnering to capture waste heat to apply to water, reducing water-heating energy requirements in homes and buildings. ML&P's new technology will reduce water consumption by as much as 75 million gallons of water annually, while AWWU's adjoining $11 million energy-recovery project is expected to save the community $1–2 million annually in energy costs.

- The **North Bay Water Reuse Authority** is committing to develop a $250 million portfolio of recycled-water and water-management infrastructure projects that will deliver a new water supply for agricultural irrigation, environmental restoration, and municipal purposes. The projects seek to capture and put to beneficial use up to 8 million gallons per year of recycled water as new supply through a diverse portfolio of projects designed to meet the needs of urban, agricultural, and environmental water users. These include water treatment using advanced filtration and UV processes, small-scale reservoirs, storage tanks, distribution systems, and groundwater-management facilities.

- **Renovate America**'s PACE (Property Assessed Clean Energy) program, HERO (Home Energy Renovation Opportunity), has facilitated $1.3 billion in financing for energy efficiency, renewable energy, and water efficiency home improvement projects in California. Today, Renovate America is setting a goal to enable $1.4 billion in new financing for an estimated 111,000 water-efficiency improvements over the next 10 years, which will help create an additional 12,000 jobs and $2.4 billion in economic impact. Through this new private investment and the projects already completed through HERO, 34.1 billion gallons of water will potentially be saved over the next 10 years.

- The **San Bernardino Municipal Water Department (SBMWD)** is announcing that in 2017, it will begin construction on Phase I of its Clean Water Factory, a $300 million recycled-water facility that will provide water for irrigation and treated water for recharging the groundwater basin that supplies the SBMWD. When the facility is completed (targeted for 2025), it will treat 6.5 billion gallons of water per year, helping to drought-proof water supplies for more than a million people.

- **Sustainable Water** is committing to deploy $500 million in capital to develop 50 eco-engineered decentralized reclamation and reuse systems requiring no upfront costs under a Water Purchase Agreement. Modeled after a campus-wide water-reclamation system used at Emory University, these systems will yield 7.5 billion gallons of recycled water annually for beneficial reuse while potentially reducing risks associated with water availability, aging infrastructure, and rising rates. The systems, which Sustainable Water will deploy in collaboration with local authorities, are designed to serve as a platform for community outreach and research while helping enable bulk water users improve resiliency and reduce their burden on existing resources.

- **Ultra Capital** will invest $1.5 billion over the next decade to support financing of decentralized and scalable water-management systems, including reclaimed water treatment, wastewater management, stormwater capture and storage, water-distribution systems, natural or biomimetic wastewater treatment, and energy converting bio-digesters. This commitment has the potential to support conservation and treatment of over 10 billion gallons of water over the next decade.

- **XPV Water Partners** is committing to invest an additional $250 million to help emerging water companies bring new solutions to the marketplace. This commitment builds on the more than $100 million to date that XPV has invested in water-technology companies.

Accelerating Development, Demonstration, and Deployment of Innovative Technologies

This past December, the Administration underscored the importance of boosting water sustainability through the greater utilization of water-efficient and water-reuse technologies; and promoting and investing in breakthrough R&D that reduces the price and energy costs of new water-supply technology. Individuals and institutions across the country responded to the Administration's call to action and are taking new steps to support these goals.

- The **Alabama Center for Rural Enterprise (ACRE)** is launching a challenge to design decentralized, on-site wastewater technologies that are sustainable, affordable, and will work with the "black belt" soils. These innovations may alleviate chronic problems of failing septic systems in rural Alabama. Nationally, over 270,000 households have experienced failures in their residential sewage-disposal systems over the last three months. The result is direct exposure of vulnerable populations to hazardous raw sewage in their home.

- The **Austin Technology Incubator (ATI),** part of the IC^2 Institute at the **University of Texas at Austin,** is launching a Water Technology Incubator (ATI Water) to accelerate the development of innovative water startups. ATI Water will build a Texas-wide network of entrepreneurs and university-based water researchers to source significant engineering and scientific breakthroughs and prove them across commercial pilot facilities. In the next five years, expects through this network to help create 500 new water-technology jobs.

- This year, **AccelerateH2O,** in partnership with **ATI Water** and **elequa,** will launch three of a planned seven regional hubs in Texas for demonstrating innovative approaches for water reuse, brackish desalination and aquifer recharge in rural communities, optimization of water systems, and smart irrigation. AccelerateH2O will work through these hubs to engage 500 youth and student "water innovators" in open-source projects, expedite commercialization of 35 early-stage breakthrough water technologies, and launch four competitions focused on addressing critical water challenges. These efforts will support AccelerateH2O's goals of increasing the recovery of "lost" water by 15–22% annually, reducing overall water use by 7.5–10% annually in drought-prone areas, and increasing sources of "new" water by 28–30% annually by 2020 across all of Texas.

- **BREW** (Business. Research. Entrepreneurship. In Wisconsin.), a water-technology accelerator program of **The Water Council,** is committing to help 75 new water-technology start-ups get their ideas launched into successful businesses over the next five years through a mentorship and intensive 6-month, strategic training program. In addition, The Water Council is announcing an expansion of its BREW Corporate Accelerator Program, with **A.**

O. Smith Corporation and **Rexnord** joining **Veolia** in the program to support start-ups in water technology.

- Led by the **City of Chicago,** the **Metropolitan Water Reclamation District of Greater Chicago,** and **World Business Chicago,** a group of prominent water entities in the Chicago region is launching Current, a new water platform connecting local public utilities, private industries, research institutions, and entrepreneurs. At launch, Current will focus on three programs: (1) a public-private research consortium; (2) a network of diverse demonstration sites for new water technologies; and (3) a circular economy/business-model innovation program to encourage public and private entities to reuse resources from wastewater and water streams. These programs are expected to support over 400 businesses and create more than $250 million in economic value over 10 years.

- The **City of Milwaukee** is announcing that the **International Water Association (IWA)** — will be establishing its first North American Regional Office in Milwaukee. In addition, the City is announcing a new formalized partnership with **The Water Council,** the **University of Wisconsin-Milwaukee,** and **Marquette University.** Together, these institutions will:
  - Build and/or attract over 75 water-focused entrepreneurs and small businesses to Milwaukee over the next five years.
  - Train up to 400 students annually with water-focused education qualifications through Milwaukee-area universities.
  - Implement the Alliance for Water Stewardship International Water Stewardship Standard at 10% of the businesses in the Milwaukee region by working with the local private sector.
  - Collaborate on practical water research, using the MetroLab framework, to address municipal and global water challenges.

- The **Clean Water Innovation Initiative**, along with its 17 founding business, government, non-profit, and research partners, is launching an EPA Water Innovation Cluster to serve the Puget Sound region in Washington. The Cluster will include three major components: (1) a physical technology-development accelerator; (2) an early (seed) stage grant/loan/equity fund to finance entrepreneurial start-ups; and (3) a national virtual network of water-industry incubators and clusters. Today, the Initiative is announcing that it will, over the next three years, work with its 30 vertically and horizontally integrated water-industry partners to support 10 companies through the accelerator and provide $2 million in seed funding.

- **Cleantech Open** is announcing the launch of its CTO-H2O water-innovation accelerator program. Through CTO-H2O, Cleantech Open and its partners will deliver commercialization training, access to capital, peer-to-peer connections, talent acquisition, water-industry expertise, and mentorship for innovative water startups. The six-month program, which expects to support approximately 30 water startups during its first year, seeks to provide national and global visibility to water technologies focused on efficiency, re-use, data, and infrastructure monitoring in the water space.

- The **Cleveland Water Alliance**, in partnership with the **Great Lakes Biomimicry Institute** and the **University of Akron,** is announcing the Water Innovation Biomimicry Program.

The program will: (1) develop and bring to market biomimicry-inspired solutions (solutions based on natural patterns and strategies) to improve water quality in watersheds; (2) prepare design guidelines, protocols, and policy frameworks to support integration of these solutions; and (3) conduct workshops on these solutions for interested private- and public-sector groups. The program's first project will utilize 3D-printing technology to bring biomimicry-inspired green bulkheads to the Cuyahoga River.

- This year, **Confluence**, a water-technology innovation cluster serving the Ohio River Valley, will hold a Technology Showcase connecting utilities with the technologists, developers, and manufacturers that can help enable specific solutions to 21 urgent water challenges identified by utilities at a conference in November 2015. Confluence will announce the Confluence "W" Prize prior to the Showcase, recognizing innovative use of technologies that address water challenges and protect public health. EPA's Environmental Technology Innovation Clusters Program works closely with EPA researchers in Cincinnati to support Confluence.

- The **Everglades Foundation** is announcing its international partnership with the **Ontario Ministry of Environment and Climate Change** on its $10 million George Barley Water Prize, a freshwater innovation challenge that seeks to find a cost-effective solution to the phosphorus-pollution problem threatening the Everglades and water bodies across the globe. This partnership aligns with recent targets made by the United States and Canada to reduce algae-feeding phosphorus entering Lake Erie by 40%. The Ontario Ministry of Environment and Climate Change has agreed to participate in the design of the challenge and in the development of cold climate parameters for testing technologies in conditions similar to the Great Lakes. Additionally, the Ministry is exploring sites and facilities to potentially host one stage of the Prize on an impacted water body in Ontario.

- **H2OTECH**, a water-technology innovation cluster headquartered at **Georgia State University** and supported by EPA's Environmental Technology Innovation Clusters Program, is announcing two efforts to grow the water-innovation economy in this area. H2OTECH, which serves 60 million people in the southeast United States, will:
  - Leverage $2.5 million per year by 2020 to support startup companies and academics for water-technology development and commercialization.
  - Expand the regional water economy job base from 30,000 to 35,000 jobs by 2020 by focusing on the emerging water-resource strategy of indirect potable water reuse in the region.

- **Imagine H2O,** a water-innovation accelerator, has launched a multiyear initiative to develop data solutions in the water industry. Today, the non-profit organization is announcing that it will expand its portfolio to source, launch, and scale 30 new water-data businesses, from monitoring and sensing to software and analytics. Imagine H2O will also double its existing partner network of utilities and companies, which provide water supply to 30 million residents across the country.

- The **International Desalination Association (IDA)** is committing to host an Energy and Environment Symposium in the United States in early 2017, bringing together approximately 250 leaders in the global desalination community to explore desalination and water-reuse issues and facilitate discussions to shape a path to a sustainable water supply for future generations.

- The **Israel-California Green-Tech Partnership** builds on California and Israel's March 2014 Memorandum of Understanding to cooperate on developing water and green technology solutions. Today, the partnership is announcing a new joint venture with the city of **Los Angeles' Cleantech Incubator (LACI)** that will culminate in the introduction of 10 Israeli companies in water, energy, and agricultural technologies to the California market. These companies will help accelerate the shift to a greener economy, with a particular focus on benefiting drought-stricken populations across the state, including the nearly 123,000 farmers in California.

- The **Massachusetts Institute of Technology (MIT) Climate CoLab**, an online crowdsourcing platform, is launching a new Energy-Water Nexus contest, soliciting high-impact proposals on the interrelated challenges of climate change, water, and energy. The challenge seeks to harness the power of collective problem-solving to catalyze innovative solutions at the water-energy nexus to build a sustainable water future.

- The **National Water Research Institute (NWRI)** is establishing a consistent validation program for innovative water-treatment technologies. Through the program, teams of experts will review the technical merits and provide a site audit of a new technology, and produce a freely accessible validation report of the technology. By providing data on the performance of new water-treatment technologies to purchasers, permitters, vendors, and the public, this program aims to significantly reduce the cost and time it takes to bring a new technology to the marketplace. In addition, later this year, NWRI—in partnership with the **WateReuse Research Foundation** (USA) and **National Centre of Excellence in Desalination** (Australia), will hold a workshop for approximately 50 delegates from the Pacific Rim nations to facilitate breakthroughs in advanced water-treatment technologies. NWRI will produce a publicly available workshop report providing a framework for funding and collaboration on advanced water-treatment technology R&D.

- **Pentair** will establish two centers over the next three years to help accelerate innovation in industrial water reuse in manufacturing, and water stewardship in food and beverage processing. Pentair expects that these centers will reach an audience of more than 100,000 stakeholders by engaging Pentair's customers and third-party organizations to innovate, validate, and collaborate. The centers will also serve to share best practices to help municipal, industrial, and commercial companies reduce their water footprint.

- The **State of Colorado** is working with private, public, and philanthropic partners to create new institutions that will help drive water innovation and infrastructure. These institutions are:
    - A Water Data and Innovation Hub in Denver to serve as a laboratory for innovation and data analytics specifically focused on water. This "Hub" is the product of a two-and-a-half day summit of private and community foundations that Colorado Governor John Hickenlooper convened in February 2016.
    - A Center of Excellence and an Intermountain Infrastructure Exchange in Colorado to (1) help leverage Federal and state funds for public-infrastructure projects with

private capital; (2) assist project proponents in considering how up-front capital can be supplied from private-sector partners; and (3) assess how project risk can be transferred to private-sector capital partners and away from the public.

- **Toray** is announcing plans to develop an R&D center in the United States to further water treatment and membrane research efforts. The center will develop research collaborations with universities and U.S. National Laboratories to further reduce the cost and energy use of desalination technologies.

- The **Toro Company** is announcing three new grants to help drive water sustainability: (1) $24,000 to support the **Wyland Foundation**'s 2016 National Mayor's Challenge for Water Conservation; (2) $6,000 per year for the next three years to support the **Western Growers Association**'s new Technology and Innovation Center and business incubator in Salinas, CA; and (3) $5,000 in training support and $3,000 in school garden drip kits as part of an educational initiative to teach children the value of water and growing food.

- The **US Water Alliance** and the **San Francisco Public Utilities Commission** are committing to convene a national commission to accelerate the adoption of urban on-site water reuse. The Commission will bring together stakeholders to develop concrete, actionable policy and regulatory recommendations for establishing standards and practices to dramatically increase the adoption of on-site water reuse in communities across America.

- **WaterNow Alliance** will recruit 100 public utilities to join the Alliance and sign on to its Statement of Principles, committing to implement sustainable-water solutions to address drought and climate change in their communities. The Alliance will work with these member utilities to substantially increase their portfolios of innovative and sustainable water solutions, with a goal of reaching at least 10 million people by the end of 2016. In addition, the Alliance is committing to work with two to three municipalities to implement sustainable water-use projects on the ground in order to demonstrate the feasibility and benefits of the projects.

- **WaterStart** (founded as Nevada's Water Center of Excellence) is a public-private partnership that has raised more than $2 million to help address the biggest technical challenges to managing water in Nevada. Today, WaterStart is announcing that **MGM Resorts** is joining the partnership to recruit, evaluate, and demonstrate new water technologies.

A key component of ensuring the adoption of groundbreaking water solutions is demonstrating that solutions are successful beyond the laboratory. Today, institutions are announcing new efforts to pilot such solutions at scale.

- **Hampton Roads Sanitation District** and its project partner **CH2M** are committing to pilot test this summer two process concepts for advanced water treatment for indirect potable reuse and several emerging technologies for mainstream wastewater treatment and nitrogen removal. These pilot studies are part of a $1 billion sustainable water-recycling initiative that will pump up to 120 million gallons of water into a coastal plain aquifer to provide a sustainable source of groundwater, inhibit saltwater intrusion, slow the rate of land subsidence in Eastern Virginia, and significantly reduce surface-water discharges from HRSD facilities into Chesapeake Bay tributaries.

- The **Iowa Department of Agriculture and Land Stewardship** and **Iowa Agriculture Water Alliance** are co-leading the Midwest Agriculture Water Quality Partnership (MAWQP), a $47M public-private partnership that leverages a $9.5M Regional Conservation Partnership Program (RCPP) award from USDA. Today, MAWQP is announcing its first major effort: the launch of a new Platform Integration Pilot, which will combine conservation and business-planning tools and environmental metrics to help farmers and agribusinesses implement conservation practices and more efficiently use resources like fertilizer, reducing nutrient loss and improving water quality and farm profitability. The partnership estimates that this pilot will reach 10,000 farmers and improve resource management on at least 50,000 acres of farmland.

- The **MIT** and the **University of Hawaii** are announcing the formation of an international team—comprised of researchers from the **Arava Institute, Jordanian German University, MIT, Technion, Tel Aviv University,** the **University of Hawaii,** and the **Weizmann Institute**—to launch a pilot study in the Red Sea region of an Advanced Pumped Hydro and Reverse Osmosis (APHRO) system to provide renewable-based energy storage and fresh water through sustainable desalination. Based on the success of the pilot, the team will explore opportunities to establish APHRO plants in other water-scarce regions, including Southern California and Hawaii.

- **Natel Energy, Inc.,** and the **University of California, Berkeley**'s Renewable and Appropriate Energy Lab are partnering to select, deploy, and assess a 1–5 megawatt project in California that will deliver 1.6 to 3.2 million gallons of groundwater recharge annually. This pilot project will help inform approaches to deliver cost-effective renewable energy while simultaneously increasing groundwater recharge, improving flood mitigation, and restoring wetland habitat.

- **River Islands,** a planned community located in the San Francisco Bay Area, is announcing a new partnership among **Nexus eWater**, local reclamation and irrigation districts, the River Islands developer, area builders, and the **City of Lathrop** to implement a system to provide River Islands with on-site water and energy recycling. By allowing homeowners to directly treat approximately two-thirds of the home's water to a quality suitable for outside irrigation, the system will reduce the intake of fresh potable water to the home.

- The **Sonoma County Water Agency**, which provides drinking water to more than 600,000 residents north of the Golden Gate, is collaborating with Federal and non-Federal partners on the Lake Mendocino Forecast Informed Reservoir Operations research and demonstration project. Working with **Army Corps**, this project will use new technologies and better weather forecasting to improve management of the Lake Mendocino Reservoir, potentially leading to water savings of roughly 10–25%.

Addressing water challenges requires understanding of the cause, scope, and impact of such challenges, as well as investigation into possible solutions. That's why today, institutions are announcing new funding and initiatives to support critical water research.

- **American Water** will invest $3 million in new research projects in 2016 to help improve water service and quality. This announcement builds on American Water's commitment to invest $5.5 billion over the next five years into needed infrastructure renewal. In addition, American Water is announcing two new collaborations: (1) with **General Electric**, to identify and explore advances in the Internet of Things to help solve pressing challenges within the water industry; and (2) with **ComEd**, to pilot an Advanced Metering Infrastructure (AMI) project that will harness new information technologies to better manage water usage and quality.

- **Arizona State University (ASU)** is launching FutureH2O, a new, five-year research initiative focused on identifying opportunities for domestic and global water security. The effort will connect ASU researchers with private- and public-sector partners to drive solutions to the most difficult water problems facing society. Specific commitments of this initiative include funding an urban landscape design and renovation campaign that reduces residential outdoor water use in at least one Phoenix metro service area by one-third by 2025; training 1,000 undergraduate and professional leaders across the U.S. Sunbelt in the next ten years to find solutions to challenges at the food-energy-water nexus; and building a food-energy-water technology test bed on the ASU campus to test innovative approaches to agriculture in the arid Southwest.

- Three universities in the Los Angeles area—the **University of Southern California,** the **University of California, Los Angeles,** and the **University of California, Riverside**—are forming a consortium to partner with the **Los Angeles Conservation Corps** and local water utilities at the SEA Lab facility in Redondo Beach, CA. The consortium will take advantage of the Lab's access to ocean water to run bench and small pilot testing on desalination, and to carry out ocean and marine studies separate from desalination. The consortium is exploring potential partnerships with the West Basin Municipal Water District and other local water agencies that are looking at ocean desalination as a water supply. The test area will be accessible to tour groups, with SEA Lab corps members demonstrating testing procedures and outcomes.

- The **Chicago Council on Global Affairs** is launching a new Global Water Policy Initiative (GWPI) to study and make recommendations on leading themes in international water policy—especially those of particular relevance to Chicago and the Midwestern United States. With quarterly workshops, in-depth research, and coordination with other major institutions involved in global water policy, the GWPI aims to engage a growing number of academics, experts, and policymakers on these issues, ranging from dozens initially to hundreds by the second year of the GWPI.

- The **City of Oceanside, CA** is launching a comprehensive pathogen study to support potable water reuse. By examining the effectiveness of upstream processes in removing and inactivating pathogens from treated wastewater, the study will provide critical data for health departments seeking to assess the safety of potable water reuse, and will help determine the level of additional treatment that reused water must undergo after passing through natural systems. The results of this study will support the City's goal of putting more than 1.6 billion gallons of recycled water annually to beneficial use.

- The **Cleveland Water Alliance,** with support from the **Coca-Cola Foundation,** is launching the second phase of its Value of Water Study. The study explores the link between clean and reliable water to regional economic growth and business, including industry, jobs/workforce development, and gross regional product. The study will tie direct, indirect, and induced economic impacts directly to regional commitments to clean water.

- This spring, the **Colorado School of Mines (Mines)** will synergistically integrate multiple water-focused research centers into the Colorado School of Mines Water Resources Institute. Through this new Institute, over 100 faculty and associated researchers will work to explore, develop, and deploy engineering and science solutions, across scales, to address water scarcity and sustainable future water supplies for people, environment, industry, and agriculture. The Institute will also help educate future water scientists, engineers, and practitioners, and help connect communities and other stakeholders to address challenges related to water and other limited resources.

- The **Electric Power Research Institute, Inc. (EPRI)** is planning to invest approximately $200 million in research and development (R&D) over the next decade to minimize the environmental impacts of water withdrawal and consumption in the electricity sector; address issues concerning the availability and cost-effectiveness of plant water-treatment options; and provide more energy-efficient and demand-responsive options for the transportation, treatment, and storage of water. In addition, EPRI is building industrywide R&D collaboratives, with participation from **DOE, NSF,** and **ARPA-E,** to evaluate the performance of a number of new, early-stage technologies that have the potential to reduce power-plant water use by 15% to nearly 100%.

- **GE** is announcing that over the next decade, it will (1) invest over $500 million into research and development to fuel innovation, expertise, and global capabilities in advanced water, wastewater, and reuse technologies; and (2) increase its customers' daily water-treatment capacity to more than seven billion gallons of water per day, up from three billion today. In addition, the **GE Foundation** is announcing that over the same time, and in partnership with **Emory University, Assist International, GE Water,** and **UNICEF,** it will support the design, installation, and training of small-scale water-purification units to produce over three billion gallons of treated water at select health facilities in developing countries. This commitment builds on the $4.7 million that the GE Foundation has already invested in this space.

- In 2014, **Georgetown University** launched coursed dedicated to helping business and diplomatic leaders understand how they can reduce the water-related risks they face and contribute to water stewardship. Today, Georgetown is announcing it will expand this training to current and future diplomats and global leaders studying in its Business School and School of Foreign Service. The university will train at least 500 additional leaders in water stewardship by 2021, and will share its educational approaches and lessons with others.

- **Kansas State University** is committing approximately $200,000 to support teams of researchers, educators, and outreach specialists working to advance scientific understanding and technology development that will minimize water use and maximize water quality in agricultural settings and at the rural/urban water interface. The teams will include 20 dedicated and 80 auxiliary researchers.

- The network of 54 **National Institutes for Water Resources (NIWR)**, in partnership with USGS, is announcing that over the next five years, approximately $18 million will be invested annually in more than 200 new, locally identified water projects. In addition, over the next five years, NIWR institutes pledge to increase strategic state, regional, and national partnerships that enhance their student-training activities, work with public and private sectors, and research addressing pertinent national and regional water issues.

- The **National Renewable Energy Laboratory (NREL)** is launching a $50,000 project to explore linked energy-water microgrids. The project, funded by NREL's Laboratory Directed Research and Development Program and conducted in collaboration with the **University of Arizona,** will explore how to improve co-management of distributed water and energy systems, with applications for remote locations around the world.

- The **Oklahoma State University (OSU)**, in collaboration with **Texas A&M University, Kansas State University,** and **USDA,** recently initiated a new project promoting the use of advanced sensor-based technologies to improve agricultural water management and minimize irrigation losses, thereby helping conserve declining agricultural water resources in the southern High Plains. Today, the collaboration is announcing that the universities and USDA will each provide over $770,000 over the next three years to develop three research sites and ten demonstration sites in collaboration with growers. In addition, the collaboration will develop new mobile apps and an online video series to assist agricultural producers, crop consultants, and government personnel in using modern sensors to increase irrigation efficiency. Finally, OSU is announcing a new research initiative on sustainable methods of augmenting limited freshwater resources for crop irrigation with produced water from oil and natural-gas exploration.

- This year, the **Pacific Institute** will address the world's pressing water challenges by: (1) working with the Business Alliance for Water and Climate Change to improve the resilience of the private sector to the greatest risks to water systems from unavoidable climate change; (2) conducting a comprehensive assessment, which will be made publicly available, of the impacts of California's severe drought on the economy and environment; and (3) expanding the efforts of the CEO Water Mandate—a corporate water-stewardship initiative administered in partnership with the UN Global Compact—to tackle water challenges facing industrial and agricultural businesses.

- **Texas A&M University**'s Water-Energy-Food (WEF) Nexus Initiative is announcing three efforts to advance awareness and understanding of, and solutions at, the WEF Nexus. The Initiative will (1) launch a community of practice that will identify and respond to national and global opportunities to assist the development of effective WEF management practices, and to develop a set of common, integrated metrics to better understand the WEF system; (2) develop an educational framework to teach stakeholders about the Nexus, through which the Initiative expects to develop 100 WEF leaders over the next five years; and (3) release comprehensive, multi-scale tools to define and quantify the interconnectivity between WEF and infrastructure. The tools will be initially deployed in the rapidly developing San Antonio, TX area and will ultimately be tested, adapted, and applied nationally.

- Researchers at the **University of California, Berkeley** and the **University of Colorado, Boulder** are launching a project to quantify the influence of vegetation and terrain snowmelt-driven runoff, which provides over half of the water supply in the western United States. By analyzing high-resolution snowpack data from the Tuolumne River Basin, the project will learn more about how this influence varies from one year to the next and from one point in the watershed to another, helping scientists to develop a new generation of predictive snowpack models and understand how climate-driven changes in vegetation may affect snowpack.

- The **University of California, Los Angeles (UCLA)** is announcing nearly $3.5 million in funding for new water-related research in alignment with the Sustainable LA Grand Challenge, a multisector research collaboration that aspires to transition Los Angeles County to 100% locally sourced water by 2050. UCLA's Water Technology Research (WaTeR) Center is announcing two new efforts to support water sustainability and security. First, the WaTeR Center will develop and test technologies for remotely monitored and controlled autonomous water-treatment and purification systems designed to serve communities and small towns, and will deploy three to four distributed water systems in remote and disadvantaged communities in California within the first two years of this project, with plans to ultimately expand to rural areas across the United States. Second, the WaTeR Center will launch new research initiatives focused on reducing the operational and energy cost of water desalination. These research projects complement UCLA's collaboration with the City of Los Angeles to construct a satellite wastewater treatment plant that is expected to provide UCLA with at least 360 million gallons per year of recycled water—supporting the university's goal to reduce potable water use per capita on campus by 36% by 2025.

- The **Prairie Research Institute** at the University of Illinois at Urbana-Champaign, in cooperation with the **National Great Rivers Research and Education Center** in Alton, IL, is launching the Resilient Watersheds Initiative, an interdisciplinary research project focused on developing coupled models to inform decision making related to water in the Illinois River Drainage Basin. The initiative will also deliver science-based education and extension services to the people living in the floodplains of the Illinois River and its major tributaries who are managing these issues. The Prairie Research Institute is planning to provide $300,000 annually to support this Initiative.

- The **University of Notre Dame's** Environmental Change Initiative and **Indiana University** are improving water quality in the Nation's heartland through the Indiana Watershed Initiative, using watershed-scale conservation to reduce nutrient runoff from farms. Today, the project team, with collaborators at **Iowa State University,** and funding from the **USDA** Natural Resources Conservation Service, **USGS, The Nature Conservancy, Walton Family Foundation, Indiana Soybean Alliance,** and **Indiana Corn Marketing Council,** is announcing the expansion of the project to include economic valuation. With farmer cooperation, the team will quantify the economic and environmental benefits of on-farm conservation to facilitate implementation of these practices across the 11 million corn and soybean acres in Indiana.

- The newly established interdisciplinary undergraduate-degree program in Water: Resources, Policy, and Management at **Virginia Tech** is designed to prepare students for rapidly expanding employment opportunities to address complex water-resources challenges for a sustainable and secure water future. Today, Virginia Tech is committing to expand this program by reaching enrollment exceeding 100 undergraduate students, increasing the program's endowment to $2 million, and expanding by 2018 to include a graduate program offering M.S. and Ph.D. degrees for students seeking advanced interdisciplinary training.

- The Leaders Innovation Forum for Technology (LIFT) is a joint initiative by the **Water Environment Research Foundation (WERF)** and the **Water Environment Federation (WEF)** to accelerate innovation in the water industry. Today, WERF/WEF are announcing the launch of a new LIFT Technology Focus Area on Water Reuse, which will establish a new network of water users identifying, evaluating, and demonstrating innovative technologies to help improve the effectiveness and reduce the costs of water reuse. WERF and WEF will collaborate with **WateReuse** on implementing the new Focus Area. In addition, in September, WEF will release a Water Reuse Roadmap to encourage resource recovery from wastewater.

- **Xylem** will help drive innovation with an intention to invest at least $300 million in water-focused research and development activities through 2018. In addition, in collaboration with the **U.S. Water Partnership** and with technical advice and input from other public and private partners, Xylem will issue a new national water-innovation challenge with funding of $50,000, focused on themes including meeting growing demand for water, protecting cities from flood and drought, and protecting the Nation's water resources. Finally, Xylem will support the efforts of the **Everglades Foundation** and the George Barley Water Prize by providing Xylem instrumentation as well as technical expertise to support field evaluations of nutrient-sensing and removal technologies.

Enhancing Data Collection, Access, and Usability

Recognizing that "you can't manage what you can't measure," institutions are announcing new efforts to enhance water-data collection, access, and usability.

- **10.10.10** is will support creation of a global center for water-data innovation. In addition, 10.10.10 is committing to launch a water-focused iteration of its "10.10.10" convenings, which will challenge entrepreneurs over the course of 10 days to solve 10 "wicked problems" in a water policy area.

- The **California Data Collaborative** is a joint effort of six California agencies to establish and accelerate the development of smart conservation targets by collectively leveraging water-use data from the 3.7 million people the agencies serve. The Collaborative is announcing that over the next six months, it will use its more than 1.8 billion records to develop a new, statewide water-conservation framework, customized to the unique needs of California's diverse communities, which these agencies will strive to implement.

- The **Community Collaborative Rain, Hail and Snow (CoCoRaHS)** network, a nationwide, volunteer precipitation-monitoring program, is announcing that later this spring, it will launch a citizen-science soil-moisture monitoring program. In collaboration with the National Integrated Drought Information System, measurements gathered through this program will be used to help validate and calibrate the increasing volumes of soil-moisture data being collected by terrestrial and satellite instrumentation worldwide.

- **Corona Environmental Consulting**'s WaterSuite™ enables real-time monitoring and management of water systems and facilitates collaboration within the water community. Today, Corona is announcing expansion of two WaterSuite™ applications. First, by the end of 2016, the Monitoring Plan Portal application, which was developed in partnership with the **State of Louisiana Department of Health and Hospitals** Drinking Water Program, will integrate more than 100,000 crowd-sourced field-sample results collected by water operators using internet-connected mobile devices. These data, and other data on the Portal, will be freely available to all 1,343 Community Water Systems in Louisiana. Second, by the end of 2016, coverage of the Source Water Protection Application will extend to 25 states and include data from more than 700 Federal, state, local, and user-specific sources; up from 13 states and more than 300 sources today. This commitment will extend active assessment and protection of watersheds supplying drinking water to more than 13.5 million customers.

- The **Desert Research Institute (DRI)** and **University of Idaho (UI)**, motivated by the White House Climate Data Initiative and in partnership with **Google**, developed ClimateEngine.org, a web application that enables users to quickly process and visualize satellite earth observations and gridded weather data for environmental monitoring and to improve early warning of drought, wildfire, and crop-failure risk. Today, DRI and UI commit to expanding ClimateEngine.org to include new drought and water-demand monitoring metrics and over 30,000 place-based averaging domains relevant for Federal and local agency rangeland, agricultural, and water-resource management in the western United States.

- The **Earth Genome** is committing $1 million over the next 12 months to build out a public data set on viable wetland-restoration opportunities in the continental United States. Wetlands are a critical component of green infrastructure to modulate freshwater supply for industry, agriculture, and municipalities. This commitment will extend a pilot developed with the **World Business Council for Sustainable Development** and in collaboration with **Dow** to assess the financial value of wetland restoration in the Brazos River Basin in Texas.

- **Los Alamos National Laboratory (LANL)** is announcing that in May, it will release new data on the impact of climate-driven heat-stress and forest mortality on Colorado River flow. The data will include basin-wide flows for multiple climate and disturbance scenarios out to the year 2100. LANL is working with multiple DOE national laboratories, Federal agencies, and large power utilities to examine these impacts on the energy-water nexus and to develop strategies for response.

- Today, **OmniEarth**, which provides technology that combines and analyzes data to determine how much water homes and businesses need, is announcing that it will expand its efforts to promote residential water conservation in California by launching commercial and agricultural water-efficiency analysis on a national scale. As part of this effort, OmniEarth will make its land-cover data available through the **California Data Collaborative.**

- **SciStarter**, a research affiliate of Arizona State University's Center for Engagement and Training in Science and Society (ASU CENTSS), is committing to advance citizen science to build a sustainable water future by:
  - Expanding the network and impact of citizen science. SciStarter has trained over 40 citizen-science teams (representing over 1,700 individuals) nationwide to take soil-moisture measurements validating data captured by NASA's Soil Moisture Active Passive satellite. Today, SciStarter is committing to train an additional 60 teams over the next 18 months—including at least one team in every U.S. state—as well as at least one trained "citizen-science ambassador" per state. SciStarter will also work with **EPA, USGS,** and **USDA** to make this citizen-science data easily discoverable and available online, and will work with **USGS t**o launch a data analysis and visualization challenge based on soil-moisture data.
  - Establishing a "Lending Library" of monitoring equipment. This year, SciStarter will launch a program in four cities—Atlanta, Philadelphia, Research Triangle Park, NC, and Tempe/Phoenix—to partner with local science museums and libraries to provide training and lend out equipment to support volunteer soil-moisture monitoring. These cities will serve as pilots for a planned collaboration among SciStarter, **ASU CENTSS,** and the **National Informal STEM Education Network (NISE Network)** to establish regional networks of lending libraries anchored in science museums across the country.

- Researchers at **Stanford University** and **Aqua Geo Frameworks** have been using a helicopter equipped with geophysical sensors to collect data on buried patches of sand, gravel, clay and water in California's drought-stricken Central Valley, down to a depth of 1,600 feet. Stanford and Aqua Geo are announcing that in April, lithologic maps based on these data of the hidden aquifers and water pathways that make up the region's poorly understood groundwater system will be made freely available for the first time. Farmers and resource managers will be able to use these lithologic maps to inform decisions about where and when to pump water or refill depleted aquifers, thus helping to ensure the long-term viability of their water supplies for domestic and agricultural uses.

- The **Gulf of Mexico Coastal Ocean Observing System,** one of eleven regional programs of the U.S. Integrated Ocean Observing System, is announcing the launch of two open data portals to monitor the health of Gulf coastal ecosystems. The Hypoxia-Nutrient Data Portal, developed in partnership with the **Gulf of Mexico Alliance,** aggregates information from multiple sources to support informed strategies needed to reduce nutrient inputs and hypoxia impacts to Gulf coastal ecosystems, extending from the inshore waters of estuaries to the continental-shelf break of the five U.S. Gulf states. The Citizen Science Data Portal aggregates data gathered by hundreds of students and citizens Gulf-wide who are monitoring environmental conditions in their local areas, allowing State, Federal, and academic programs to supplement their datasets with much more granular, localized information.

- **The Water Council,** working with the Innovation Exchange, has launched the Global Water Port, an online research tool which enables access to thousands of real-time water-data sources. Today, the Water Council is partnering with the **Federal Lab Consortium (FLC)** and the **U.S. Water Partnership** to make data from Federal labs more accessible through the Global Water Port.

- The **University of Alaska Anchorage,** as part of the EPA-funded **DeRisk Center,** is developing a GIS application that makes the data housed within the EPA Safe Drinking Water Information System database usable for a water-plant operator. The app's goal will be to allow an operator to easily determine where else in their state a particular water-treatment process is being used, thus reducing the steps and cost required in seeking process assistance. Today, UAA is committing to complete the app and make it publicly available through the Center's website by summer 2017, and to share the app at state, national, and international conferences throughout its development.

- The **University of California Water Security and Sustainability Research Initiative** is committing to develop, by 2018, an integrated water-accounting system. The system will include a new basis for managing groundwater by using a novel combination of conventional groundwater-level data and modeling tools that will be disseminated to hundreds of water managers by 2017, including those in 127 California state-defined groundwater basins.

- Over the next 18 months, **Water Canary** will launch a water-quality data collection service, offering real-time nutrient data collected from sensors the company installs and maintains to make it affordable for businesses, farmers, scientists, and government agencies to use water more efficiently and eliminate the waste of excess fertilizer in agriculture. The company has set the goal of bringing all major river systems in the continental United States online by 2020, increasing the total number of publicly available real-time data points from under 5 million a year today to over 10 billion.

- The **Water Funder Initiative** is launching Project Water Data. Project Water Data is an effort to work with Federal, state, and local governments, as well as private- and social-sector partners, to modernize data systems that support healthy communities, thriving agricultural systems, and clean waterways for our wildlife. With seed funding provided by philanthropic partners in the Water Funder Initiative, and support from the **Association of California Water Agencies,** the **City of Los Angeles,** the **Colorado Water Conservation Board, Environmental Defense Fund, DC Water,** the **Metropolitan Water District of Southern California, Milwaukee Metropolitan Sewerage District, Northern California Water Association,** the **State and Federal Contractors Water Association, The Nature Conservancy, Trout Unlimited,** and others like the **Aspen Institute,** Project Water Data will: (1) mobilize a coalition to highlight the value of open, integrated water data in supporting better decision-making and citizen engagement; (2) develop a core set of principles for open and integrated water-data systems; and (3) unlock water data from all sectors and develop products that increase the discoverability, usability, and interoperability of data.

- **WaterSmart Software** will, over the next ten years, expand its utility partnership community to reach more than 45 states and over 5,000 water utilities with its digital customer-engagement technologies. The company estimates that as a result of this expansion, more than 128 billion gallons of water (the equivalent annual water use of 1.2 million households) will be saved, and carbon-equivalent emissions will be reduced by more than 1 million metric tons. In an effort to advance understanding of water use and related energy consumption and impact on greenhouse-gas output and global climate change, WaterSmart is committing to make newly available aggregated and anonymized water-consumption and demographic data from its operations freely available in a standardized format for public-research purposes.

Conserving Water and Watersheds

Reducing water use and maintaining the integrity of our Nation's natural systems are two important parts of ensuring that everyone—humans, animals, plants, and ecosystems—have access to sufficient water when and where they need it. Individuals and institutions responded to the White House call to action with new steps to conserve water and water basins across the United States.

- The **Alliance for Water Stewardship (AWS)-North America**, a program of **The Water Council,** is promoting corporate water stewardship in the United States through implementation of the AWS International Water Stewardship Standard (AWS Standard) across U.S. industrial and agricultural sectors. Today, AWS is committing to working with 200 large industrial and agricultural water-using sites to implement the AWS Standard, to provide a framework to help sites use water more strategically and identify and mitigate internal and external water-related risks. AWS expects this effort to save more than one billion gallons of freshwater over the next decade.

- At the **Society of Freshwater Science** annual meeting in May 2016, **American Rivers** will host a special session entitled "Rivers at Risk," at which distinguished scientists in the field will assess threats to rivers in the 21st century and launch a series of papers and a special journal issue leading up to the 50th anniversary of the Wild and Scenic Rivers Act on October 18, 2018.

- The **Beverage Industry Environmental Roundtable (BIER)** will this year launch a "Project Cost Curve Database" that will collect information on projects related to energy and water conservation at BIER's 20 global beverage-company members and their 1800 facilities. BIER members and facilities will be able to use this information to inform the development of their own resource-conservation projects. In addition, BIER is launching "Future Scenarios 2025", an integrated effort to explore how access to water and other resources is likely to change over the next decade, and to introduce business leaders and other stakeholders in the beverage industry to practices that could increase water sustainability.

- Under the leadership of the **Bonneville Environmental Foundation**, the business water-stewardship networks **Change the Course** and **Protect the Flows** are merging. The new organization is announcing that over the next 18 months, it will work to educate and engage 500,000 Americans in water-conservation practices, and will provide an award to a network business member for innovation and water conservation in the West.

- **Coca-Cola** will work to establish more than 10 corporate partnerships by 2017 to expand its efforts to engage with local communities, government, and business partners to support the sustainability of local watersheds in the United States.

- **Cox Enterprises** has set goals of becoming water- and carbon-neutral and sending zero waste to landfill by 2044. As part of these goals, Cox is committing to reduce its water use by 6.5% annually through four parallel efforts: (1) reclaim approximately 10 million gallons of water annually through Water Conservation Centers; (2) partner with American Rivers and Ocean Conservancy for water cleanups in Cox locations across the Nation; (3) deploy new technology and water-efficient fixtures throughout Cox facilities; and (4) use xeriscaping at locations with large land footprints to save more than 42 million gallons of water annually.

- In response to the ongoing western drought, the **East Bay Regional Park District** in California, the largest regional park system in the United States, has established a goal to save 250 million gallons of water over the next five years. As part of this effort, the Park District is announcing that it will (1) eliminate standard grass in some high-use areas (replacing with drought-tolerant grasses), and partner with local sod growers and seed companies to make drought-tolerant grasses available to the general public; and (2) convert a number of grass areas into native plant gardens for water-efficiency and public-education purposes. These two initiatives will potentially cut water use by 30%–50%.

- **Ecolab** is committing to improve water-productivity intensity by 30% in its United State facilities by 2020 (from a 2015 baseline). This will conserve approximately 100 million gallons of water over five years.

- The **Environmental Defense Fund (EDF)** is announcing a partnership with **Pecan Street, Inc.** to gather data and conduct analysis to help 50 households in Houston and Austin understand the connection between their water and energy use. Results of this analysis will help Houston and Austin reduce the water and energy footprint of the more than 3 million utility customers in the two cities. In addition, EDF will work with the **University of Texas at Austin,** the **University of Texas at San Antonio,** and the **University of California, Davis** to help water providers of three to five major Texas cities better manage the energy use embedded in their water systems, with an additional one or two states to be announced later this year.

- **General Mills** is committing to champion development of stewardship plans by 2025 for high-risk watersheds in its global supply chain. As part of this commitment, General Mills will work to develop the science and tools necessary to achieve sustainable groundwater management in relevant California watersheds, working in partnership with stakeholders in each watershed, **The Nature Conservancy,** and **Sustainable Conservation.**

- To minimize water use in apparel manufacturing, **Levi Strauss & Co.** has created Water<Less™ finishing techniques, which save up to 96% of the water typically used in the denim-finishing process. Today, Levi Strauss & Co. is committing to produce 80% of its products using its Water<Less™ techniques by 2020, and to make its Water<Less™ methods, along with other water-conservation approaches and tools, publicly available to others within and outside the apparel industry. In addition, Levi Strauss & Co. is committing to, by 2020, use 100% sustainable cotton from sources such as Better Cotton and recycled cotton that use less water and fewer pesticides.

- **Nestlé** is announcing that it will implement the Alliance for Water Stewardship standard in its California operations. Additionally, by 2017, Nestlé will put in place operational plans that will save 144 million gallons of water annually in Nestlé's California factories, building on its commitment to transform one of its nine California factories into a "zero water" facility.

- The **Sonoran Institute**, in partnership with **the Central Arizona Conservation Alliance,** is announcing an initiative to develop a collaborative conservation plan for the Phoenix metropolitan area. The plan will help the 24 cities and towns in Maricopa County protect local watersheds and encourage sustainable recharge of aquifers, and thereby fulfill state requirements for water availability needed to permit future growth. Wide-spread adoption of this plan will support the long-term supply of groundwater needed to sustain the economic vitality of the region while ensuring the conservation of the region's ecology.

- The **Texas Environmental Flows Initiative**, a joint effort of the **Meadows Center for Water and the Environment, Harte Research Institute, National Wildlife Federation, The Nature Conservancy, Ducks Unlimited,** and the **National Fish and Wildlife Foundation** is committing to the development of the foundational science and market analysis to launch a water-transaction market in Texas for the benefit of bays and estuaries. Over the next two years, the Initiative will execute at least one significant water transaction with demonstrable benefit to ecological resources injured by the Deepwater Horizon oil spill, and lay the groundwork for market development in three bay systems whose ecological health and commercial fishing productivity are imperiled by declining freshwater inflows.

- **Trout Unlimited (TU)** is announcing two new efforts to improve drought resilience in two river basins in the western United States. First, in the upper Green River flowing through Wyoming, TU will work with Wyoming cattle ranchers to identify workable approaches for ranchers to share water with downstream municipalities. This effort will potentially save approximately 1.5 million gallons of consumptive water use per year. Second, TU will finish work in 2016 with the **Methow Valley Irrigation District** in eastern Washington State, investing in aging water infrastructure in a way that increases the reliability of their irrigation-water delivery, sends water to the town of Twisp, and increases flows for imperiled salmon and steelhead.

- With support from the **William Penn Foundation,** the **University of Delaware** and **Nature Conservancy of Delaware** are announcing the formation of the Brandywine Christina Healthy Watershed Fund in Delaware and Pennsylvania. The water fund is designed to provide up to $5 million per year from public-private water utilities and other public-private sources for agricultural conservation programs upstream in the regional watershed that provides drinking water to over a half million people in both Delaware and Pennsylvania.

- **Water Quality Indiana (WQI)** is launching a three-year river-restoration and conservation initiative that will engage more than 50 high-school seniors and undergraduates from private and public universities in east-central Indiana. With 2016 funding of $28,000 from the Virginia B. Ball Center for Creative Inquiry at **Ball State University,** this inter-institutional partnership will develop a virtual space for student-contributed water quality data and scholarship shared through learning modules and multimedia. Core aspects of this initiative will be disseminated to over 250 educators through national-level workshops and replicated within a growing network of partner institutions.

- The **World Wildlife Fund**, together with bi-national partners including **The Coca-Cola Company, Tecnológico de Monterrey, Desert Landscape Conservation Cooperative,** and **South Central Climate Science Center,** is announcing a Rio Grande/Rio Bravo Forum to be held in February 2017 to share successes and identify innovative solutions to address the region's mounting water stress. The Forum will bring together over 100 diverse water users from the United States—building towards a Binational Forum including both the United States and Mexico—to cooperate on ensuring the long-term integrity of the Rio Grande/Rio Bravo basin, which provides fresh water for over 13 million people.

Helping Communities in Need

Access to clean, safe drinking water is often a particular problem for predominantly poor, minority, or rural communities. Today's announcements include responses to the White House call to action that will target water assistance to those in need.

- The **Dow Chemical Company** is partnering with **Genesee County Habitat for Humanity** to offer free water-filtration systems to 150 Habitat for Humanity homes in Flint, MI. Through this partnership, Dow will provide the reverse osmosis (RO) technology for the the water-filtration systems that will be installed in residents' homes.

- **Evoqua** will donate 10 Sky Hydrant water-filtration units—each with the capacity to meet the daily water needs of more than 6,000 people—to underserved, emergency, and disaster-relief efforts in the United States. In addition, Evoqua is committing to (1) invest an additional $50 million in research and development to further expand water reuse and reclamation efforts across municipal and industrial applications in the United States; and (2) to, by 2021, increase the amount of water the company treats for reuse and reclamation to 5 billion gallons of water a day—double Evoqua's current capacity.

- **Micronic Technologies** is announcing that it will provide its MicroDesal™ technology at reduced cost to small community water/wastewater facilities to moderate deteriorating infrastructure. Micronic is also committing to developing this technology through partner collaborations—in particular, with the **University of Virginia**'s College at Wise—to provide secure, safe, potable water to small, remote communities throughout the United States and the world.

- In response to extreme drought in the State of California, the **San Francisco Foundation** is investing $150,000 in partner organizations to address social vulnerability and build community resilience to water scarcity in low-income communities and communities of color. The grants will help community groups engage in the implementation of California's new drought measure and ensure that the associated public revenues build sustainable water projects in disadvantaged communities, among other activities.

- **Triple Clear Water Solutions, Inc.,** a company that provides plug-and-play water-purification technologies, is committing 1% of its sales—which is expected to be more than $1 million over the next several years—to fund clean-water initiatives in communities in need of help.

- **WaterFX** and **Partners in Health** have teamed up to form OpenWATER, a non-profit organization dedicated to accelerating the deployment of innovative water technologies for enhancing water security in resource-poor and underrepresented communities such as rural communities, tribal nations, and island territories. OpenWATER will draw on WaterFX's experience in sustainable water treatment—to deliver water technologies in tens of communities over the next two to three years.

Raising Public Awareness

With water often traveling hundreds of miles before flowing out of a tap, many Americans don't know where their water comes from, the underlying stresses facing their water supply, or approaches that can be adopted to help ensure long-term water security. Individuals and institutions are today announcing new efforts in response to our call to action to raise awareness and improve knowledge of water in the United States.

- This year, **America's Watershed Initiative (AWI)** will launch a #raisethegrade communications campaign on the importance of the Mississippi River watershed, which received a D+ in the most recent AWI Report Card. The campaign will focus outreach to key private- and public-sector leaders in the watershed, including representatives of 400 institutions that participated in the Report Card process. AWI will pair this leader outreach with a mass-media strategy targeting all five basins of the Mississippi River watershed.

- **Blue Legacy, Global Water Challenge,** the **U.S. Water Partnership, Veolia** and partners are launching SOURCE, a new, freely accessible water-storytelling platform that will educate the public about the importance of water and offer recommended water-management practices.

- A diverse coalition of global businesses with significant supply chains or operations in California are announcing their commitment to join *Connect the Drops*, a campaign urging policy measures by decision-makers to maximize California's local and state water resources. Launched by **Ceres**, a nonprofit sustainability advocacy organization, the campaign has 23 signatories from the private sector. The signatories joining the campaign today — Anheuser-Busch InBev, Annie's, Eileen Fisher, Kellogg Company, and Xylem — have collectively committed to saving nearly 1 billion gallons of water through 2020 through current efforts and new goals.

- This spring, **GRACE Communications Foundation** is releasing a Spanish version of its Water Footprint Calculator, a free, nationally used tool that illustrates how everyday actions impact water use. The new release will allow the tool to help an even wider audience understand their water use and reduce their water footprint.

- For the past eight years, **Green Schools** has been honoring environmental excellence, innovation, and stewardship across the United States through its Annual Green Difference Awards program, which is sponsored this year by the **Walmart Foundation.** This year, Green Schools will add a category for Best Practices in Water Innovation. In addition, Green Schools is announcing a new Water Innovation Challenge. K-12 students will compete in three areas — Best Green Schools Water Practices, Best Student-led Water Practices in our Community, and Best Innovative Water Business Idea — and will have the opportunity to pitch their ideas to leading businesses and government officials at the 9th Annual Green Schools Summit.

- The **Interstate Council on Water Policy (ICWP)** is announcing that over the next 24 months, it will engage over 100 partners — including State water-management agencies, interstate basin commissions, and NGOs — in (1) communicating the importance of water data and science in informing water resources policy and planning; and (2) identifying and promoting opportunities to enhance State and regional water-resource planning efforts. As part of this effort, ICWP will convene at least two national stakeholder workshops focused on water-data collection and water use, and at least two workshops highlighting successful practices from State and interstate water-plan development.

- The **Irrigation Association** and the **National Ground Water Association** are partnering to launch a new educational campaign aimed at helping the Nation's 121,000 farms enhance water efficiency and reduce energy consumption of their 476,000 irrigation wells. The campaign will include an online resources portal; a national series of presentations, seminars, and webinars; and collaborations with USDA's Natural Resources Conservation Service and extension office.

- **Itron** has partnered with Professor Michael Webber at the **University of Texas at Austin** to create and distribute an interactive curriculum that teaches key concepts about water and energy for K-12 students, colleges, industry, and the general public. This curriculum will combine traditional content with multimedia components such as audio and video, along with dozens of interactive exercises, maps, and games in order to improve water and energy literacy, encourage conservation and resourcefulness, and inspire the next generation of innovators. Working with its community partners nationwide, Itron will make the app-based curriculum available free of charge, with a goal of reaching at least 10,000 students in 2016 and expanding globally in 2017.

- **Levi Strauss & Co.** will expand its partnership with the **Project WET Foundation** to train Levi Strauss & Co. employees to become water-conservation ambassadors, empowered to educate their local communities about the importance of saving water. Together, Levi Strauss & Co. and Project WET have developed a training curriculum to teach employees and local communities about the impact their clothing has on the planet, and changes individuals can make in their daily lives to conserve water. Levi Strauss & Co. is committing to provide this water-education training to 100% of the company's corporate employees by 2020.

- The **National Environmental Health Association (NEHA)** is announcing an initiative to increase awareness of (1) the effect of the environment on water supplies and the role environmental-health professionals play in keeping water safe; and (2) approaches to ensure that water reuse systems do not negatively impact public health. Under this initiative, NEHA will work with its members and 50 affiliate organizations to compile information on these topics, and share the information collected with partner organizations, environmental-education programs, and local health departments.

- The **Smithsonian**'s Museum on Main Street program will initiate a traveling educational exhibition, *Water/Ways*, which explores the centrality of water in our lives as an environmental necessity and an important cultural element. Beginning in May 2016, the Smithsonian, in partnership with state humanities councils, will launch the exhibition on simultaneous year-long tours of five states—Florida, Idaho, Illinois, Minnesota, and Wyoming—and will also support the exhibition traveling to more than 180 towns across 30 states over the next six years. The exhibition will be accompanied by the launch in June 2016 of a website that will serve as a gateway to share resources and collect stories on water.

- Over 1.6 million people in the United States still lack basic plumbing facilities such as a toilet, a shower or bathtub, and running water. **Southeast Rural Community Assistance Project, Inc. (SERCAP),** working with the **National RCAP,** is committed to reducing this number by 10% over the next ten years. Through a combination of social-media campaigning, crowd-funding, and its "A Day Without Indoor Plumbing Challenge", SERCAP will bring awareness of this problem to the general public, in addition to the private and government sectors.

- The **Theodore Roosevelt Conservation Partnership (TRCP)** is announcing that more than 800 sportsmen and women have signed a petition recognizing serious risks to the American water supply, including fish and wildlife habitat, and calling for action to reduce the risks of water shortages, create flexibility in water management, and improve the reliability of water systems on a basin scale.

- **ThinkWater** is a national campaign supported by **USDA**'s National Institute of Food and Agriculture and led by the **University of Wisconsin-Extension** and **Cabrera Research Lab** to help people think differently, and care more deeply, about water. Today, ThinkWater is committing to work over the next two years to build a national coalition of at least six state-based networks to engage water researchers, educators, and extension agents in solving water-related problems through better systems thinking. This effort will begin this spring with the Wisconsin Water Thinkers Network. ThinkWater expects to directly engage approximately 200 experts and practitioners directly involved with water issues through the Wisconsin network, and approximately 1200 once all networks in the coalition have been established. In addition, ThinkWater is announcing that this spring it will launch "Systems Thinking Made Simple," a free, interactive online course designed to introduce systems-thinking concepts, tools, and resources to water researchers, educators, extension agents, and citizens across the country.

- **Water Education of Latino Leaders (WELL)** is launching WELL 2.0, a new effort to provide basic education about water infrastructure and finance to municipal officials in California. WELL 2.0 is a series of fully-funded conferences, roundtables, and gatherings that bring special-district elected directors into contact with municipal elected leaders to enable local communities to overcome water challenges and ensure that safe, sufficient, and affordable drinking water is available to poor and underrepresented minority communities.

- Through their water-focused corporate citizenship initiative Watermark, **Xylem** has set a goal of logging 100,000 hours of employee volunteer time over the next three years in projects to include presentations and hands-on water-monitoring activities at local schools and community centers, water source clean-up activities to protect local water resources, and charity "Walks for Water" to raise funds and awareness of water issues.

- **Zurn Industries LLC** will provide water-efficiency training to 1,000 municipal agencies and utilities as well as 10,000 building owners, architects, engineers, and contractors. The training will be focused on reducing water use through water-efficient products and practices with the goal of saving 114 billion gallons of water over the next decade.

Delivering Tools and Resources

Today's announcements include the release of a broad range of tools and resources to support individuals, communities, and governments of all levels in developing and implementing solutions to key water challenges in the United States.

- This year, the **Alliance for Global Water Adaptation (AGWA)** is publishing the Climate Risk Informed Decision Analysis (CRIDA) methodology, a guidance document that seeks to enable water managers to plan for and manage water resources sustainably over decades and centuries despite deep future climate uncertainty. Also this year, AGWA will launch a global community of practice based on the publication, along with a series of graduate and post-professional courses offered initially through universities in the United States and Europe.

- The **American Water Works Association** is releasing Water Loss Audit Software version 5.0, a free tool that has been updated to support audits for water systems of all sizes. AWWA is challenging 1,000 water utilities to complete a water audit using AWWA's newest software in the next two years and report their findings on AWWA's website.

- The **Association of Metropolitan Water Agencies (AMWA),** together with **EPA,** the **Association of Clean Water Administrators,** the **American Public Works Association,** the **Association of State Drinking Water Administrators,** the **American Water Works**

Association, the **National Association of Clean Water Agencies,** the **National Association of Water Companies,** and the **Water Environment Federation,** is releasing important updates to the Effective Utility Management (EUM) and the Keys to Management Success, a framework for sustainable water-utility management. The updates incorporate new science and approaches in water-utility management, such as performance monitoring and expanded use of data from automated and smart systems to optimize operations and minimize water loss.

- This spring, the **Climate Registry** will begin a pilot of the Water-Energy Greenhouse Gas (GHG) Guidance with water agencies located in Southern California Edison's territory. This pilot will serve to operationalize the Guidance as a resource for water agencies to potentially improve their water-, energy-, and GHG-reduction abilities with better data.

- **DC Water** and the **Water Environment Federation** are developing a National Green Infrastructure Certification Program to promote a skilled green infrastructure (GI)-workforce and help support community-based job creation in U.S. cities. The Program will provide certifications to individuals performing the installation, inspection, and maintenance of GI as having the required knowledge, skills, and abilities to support long-term performance and sustainability of GI systems, which can help reduce combined sewer overflows and provide triple-bottom-line benefits. The **Milwaukee Metropolitan Sewerage District** is joining with **WEF** to help advance the certification.

- The NSF-funded **Engineering Research Center for Re-inventing the Nation's Urban Water Infrastructure (ReNUWIt),** a research partnership among **University of California, Berkeley, Colorado School of Mines, New Mexico State University,** and **Stanford University,** will seek to advance urban water governance by releasing a set of decision-support tools this year that will allow utilities to quantify regional urban water resiliency and sustainability; promote the diversification of urban water-supply portfolios by enabling virtual trading in regions with shared water resources; and support integrated management of water-reuse and stormwater-recharge systems. These tools are being developed in collaboration and partnership with water and wastewater utilities in California and Colorado, and will be tested by utilities and regional planning agencies.

- The **Family Farm Alliance** is releasing a report compiling case studies in several states (CA, CO, NM, OR, WA, and WY) that highlight real-world examples of water conservation, water transfers and markets, aging infrastructure problems, watershed restoration, and ecosystem enhancement on farms and ranches. The report will describe unique complications facing local water users and creative solutions, helping to scale efforts that support better management of water for both economic purposes and environmental uses.

- The **Global Lake Ecological Observatory Network (GLEON)** and the **North American Lake Management Society (NALMS)** are announcing new resources to allow broader participation in lake- and water-quality monitoring. In partnership with the U.S. Geological Survey, Esri, and other institutions, GLEON will further develop the Lake Observer mobile application with new mapping and data-visualization features to help researchers and citizen-scientists record lake and water-quality observations. GLEON and NALMS will also make the app available for use in the annual Secchi Dip-in event and partner with the

EPA to make collected data publicly available for the first time via the Water Quality Portal. In addition, NALMS is working with graduate students to develop online video tutorials to help interested students participate in water-quality monitoring and lake management. The tutorials will be posted online this summer.

- The **International Association of Plumbing and Mechanical Officials** is releasing *The Drought Toolkit: A Community Guide to Achieving Water Efficiency Today.* The freely available toolkit will help diverse stakeholders—including city councils, local planning and development departments, code-enforcement officials, state construction boards, and state legislatures—realize more than 20% in water savings in the built environment.

- **Mammoth Trading** is launching two new trading markets for water leasing in over 500,000 acres of irrigated farmland: (1) for groundwater trading in western Nebraska; and (2) for surface-water trading in central Washington State. These markets, which will be set up for municipalities and communities for free, will seek to leverage the power of computer optimization to automate the process of checking complex regulatory rules for trading and to generate economic gains among participants. By monetizing the value of conserved water, water leases generate a potential new revenue for water users and reward innovation in water use at the farm level. Mammoth Trading grew out of NSF- and USDA-funded research, which was commercialized through the NSF Innovation Corps.

- In 2017, the **MIT Center for Advanced Urbanism (CAU),** with funding from the MIT Abdul Latif Jameel World Water and Food Security Lab, will release new design guidelines for building constructed wetlands to enhance urban water resiliency and flood protection through stormwater capture in cities. These guidelines incorporate a unique combination of engineering, design, and ecological systems to rethink how the natural landscape, water infrastructure, resiliency, and urbanism come together. This effort builds on CAU's New Meadowlands proposal to address flooding in New Jersey.

- In 2016, the **NELAC Institute** will release a new standard to improve the competency of laboratories that conduct testing on drinking, wastewater, and surface water. This standard will build upon existing standards regarding how laboratories measure and report contaminants in water at low concentrations.

- The **North American Lake Management Society (NALMS)** is announcing the addition of a new track to its Certified Lake Manager (CLM) and Certified Lake Professional (CLP) programs, allowing the participation of undergraduate and graduate students. This new track will help grow the water workforce by (1) giving students an opportunity to gain skills needed to become proficient water-quality professionals; (2) increasing the number of certified CLMs and CLPs; and (3) expanding the network of people with the knowledge of basic freshwater science, ecology, and other areas needed for informed lake management.

- In July 2015, **Rancho California Water District (Rancho)** launched MyWaterTracker, a digital platform that enables water users to visually see and track their water use on a day-by-day basis and compare current water consumption to individual household water budgets. To date, Rancho reports that use of the tool has resulted in District-wide water savings of 30% over 2013, or enough to serve approximately 20,000 households. Today,

Rancho is announcing that it will launch a mobile-app version of this tool in summer 2016, which will include additional hourly water-use data and leak alerts and is expected to reach over 33,000 residential and agricultural customers.

- **Texas A&M University (TAMU)** will develop and implement a web-based technology that provides real-time water-usage information directly to water-utility customers, empowering them to make more informed decisions about their water consumption. TAMU is partnering with the Texas cities of Arlington and Round Rock to expand use of the technology, and will continue to partner with other water utilities to test and refine the system.

- **The Nature Conservancy**, in partnership with the **American Planning Association**, the **Association of State Floodplain Managers**, the **National Association of Counties**, and **Sasaki Associates**, will develop a free, publicly available online siting guide that communities can use to identify a suite of potential nature-based solutions to flooding challenges. The guide will serve as a helpful tool to support municipalities in investing in natural systems and nature-based solutions to address their flooding challenges.

- The **Water-Culture Institute**, in collaboration with **University of Arizona** and the **Southwest Water Technology Cluster,** is developing an "Ethics-Based Decision Support Tool" (EBDST) for guiding technology, policy, and investment decisions in the water sector. Each EBDST shares a common framework, which is tailored to specific users through community engagement to set value priorities for water decision-making. The resulting decision tool can be incorporated into existing water-governance arrangements. The EBDST will be piloted in Santa Fe, NM and three other cities in 2016 and 2017, and scaled nationwide beginning in 2018.